Work

Energy

and

Power

© sciencebod 2014

Preface

Work, energy and power are Physics concepts that are often misconstrued by young students. In this book, we try to present the meanings of these concepts in coherent and yet easy-to-understand style. Numerical problems have also been used to consolidate mastery and usage of the concepts, especially to persons at the early stages of scientific study.

WORK, ENERGY AND POWER

Introduction

1

The concepts work, energy and power are three very important concepts in physics that we encounter in our everyday life. Very often we hear people make statements like the following: I am going to work, I can't do that work because I don't have the energy required (I am tired), that man is very powerful, that machine is very powerful. Therefore all we are going to do in this chapter is to give the physics explanations of these concepts (i.e. what these statements mean in physics).

1.1 The concept of work

2

In physics, work has a precise definition that differs from our everyday usage. Work is said to be done whenever a force moves its point of application through a distance in the direction of the force. The amount of work done is equal to the product of force and the distance moved in the direction of that force.

i.e. work = force × distance in the direction of force.

$$W = F \times S \text{ ------------------------1.1}$$

More formally, work is defined as the product of the force and the displacement in the direction of the force.

1.2 What is the S.I. unit of work?

The S.I unit of work is Joule (J) or Newton-metre (Nm)

1J is therefore the same as 1Nm, and they are both equivalent to $1kgm^2s^{-2}$
$(1J = 1Nm = 1kgm^2s^{-2})$

One Joule is the work done when a force of 1 Newton moves (the point of its application) a distance or displacement of one metre.

1.3 Illustration of work done

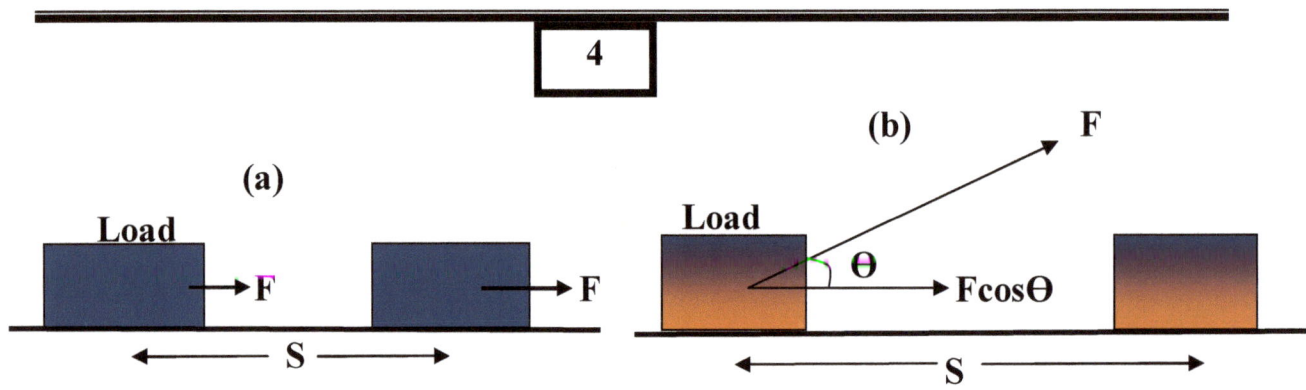

Fig.1.1

In figure 1.1(a), the work done by the force F in moving the load a distance S is given by:

$$\text{work} = F.S$$

But in figure 1.1(b), the work done is give by:

$$\text{work} = F\cos\theta \times S ----1.2$$

where $F\cos\theta$ is the horizontal component of F

This is because in figure 1.1(b), the force has moved the load a distance S in the horizontal direction, but the force itself is not acting horizontally (it is acting at an angle ϴ to the horizontal), and so we must resolve the force horizontally to make sure it is in the same direction as the distance S.

This emphasizes the last phrase in our definition of work: work is the product of the force and the displacement <u>in the direction of the force</u>.

For figure 1.1(a), the force F and distance S are already in the same direction, so we needed not to do any resolving of the force. If we did, we'll still find that the value of the force will never change.
This is because, the angle ϴ between the force and the distance is 0 in this case. Therefore FcosϴF will be Fcos0, which is the same as F (since cos0=1).

Note Also that if the force is perpendicular to the horizontal,...

Then the force can't move the load horizontally, and the work done by this force in moving the load horizontally will always be zero.

If (in the illustration above) the force is perpendicular to the horizontal (i.e. ϴ = 90^0), then Cosϴ = $Cos90^0$ = 0.

And so, Work = $F \times cos90^0$ = F×0 = 0.

displacement
(a)
(b)

Fig.1.2(a) before resolving F horizontally, (b) after resolving F horizontally

Let's now look at a first problem

A box of mass 40kg is being dragged along the floor by a rope inclined at 60^0 to the horizontal. The frictional force between the box and the floor is 100N and the tension on the rope is 300N. How much work is done in dragging the box through a distance of 4m?

(A) 680J (B) 400J (C) 200J (D) 100J   **(JAMB)**

Solution

Solution:

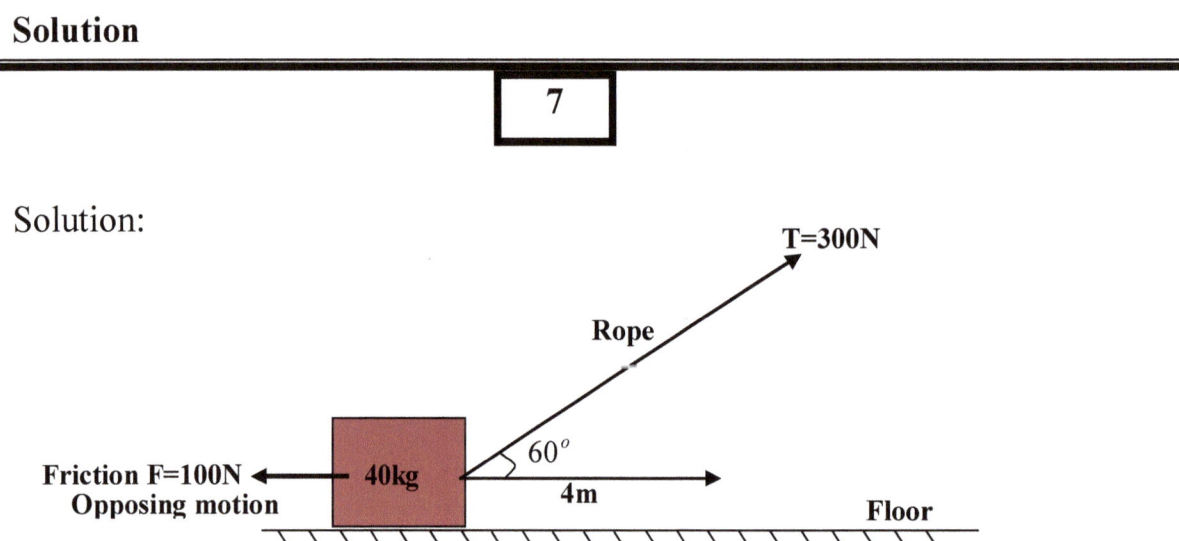

(Recall, Tension is the name given to a force acting on a rope or string/spring)

Since the distance is moved horizontally, we have to resolve the tension T horizontally, and that gives $T\cos\Theta = 300 \times \cos60 = 150N$.

Next, we remember that the 100N frictional force opposes motion, and so we have to deduct it from the horizontal component of T, and that gives $150 - 100 = 50N$ as the resultant horizontal force acting on the box.

Therefore, the work done is $F \times S = 50N \times 4m = 200J$

Question 2

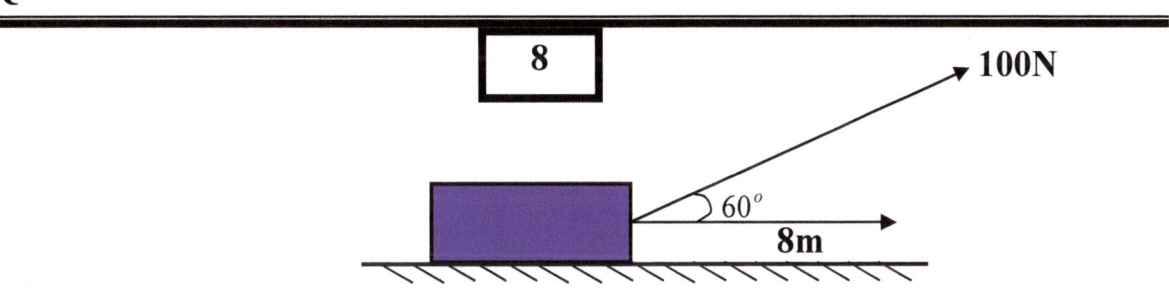

In the figure above, the work done by the force of 100N inclined at an angle of 60^0 to the object dragged horizontally to a distance of 8m is

(A) 800J   (B) 600J   (C) 100J   (D) 400J   **(JAMB)**

You should be able to do that on your own

9

Solution:

Work done = $F\cos\Theta$ × distance

(Where F = 100N, $\Theta = 60^0$ and distance = 8m)

$\Rightarrow$ Work done = $100N \times \cos60 \times 8m = 400J$

1.4 Work done in a gravitational field

A gravitational field that we all are familiar with is the Earth's gravitational field.

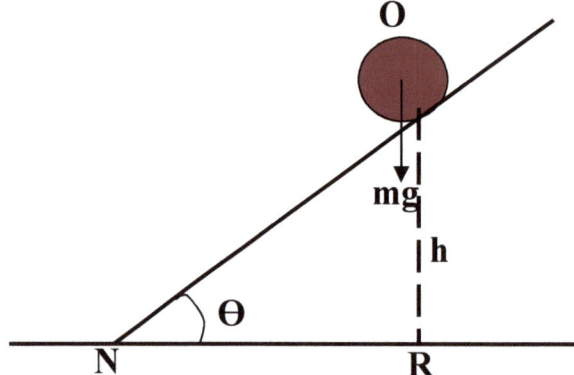

Fig 1.3

Suppose a body of mass m is released from a point O to fall vertically through the path OR as shown in figure 1.3, then the work done is

W = F × S   = mgh

(where the force F in this case is the weight mg of the body, and h is the distance through which it has to fall)

On the other hand, if the body was to fall along the smooth plane ON inclined at an angle Θ to the horizontal, then we have to resolve the weight mg of the body to get its component along ON.
Doing this, we'll get mgSinΘ (see our book chapter on 'Scalars and Vectors' for how this is done)

From trigonometry, $sin\theta = \dfrac{h}{ON}$

therefore, the distance ON $= \dfrac{h}{sin\theta}$

The work done by gravity when the body falls along the plane ON is therefore:

F×S = mgSinΘ $\times \dfrac{h}{sin\theta}$ = mgh

Gravity is conservative!

We see from plan 10 above that the work done by gravity when the body falls through either paths (OR or ON) is the same.

This clearly shows that the work done by gravitational force depends only on the vertical distance moved, but not in the path taken.

It also goes to confirm that gravitational field is a conservative field. Work done in such a field is recoverable without a loss.

Question 3

An object of mass 50kg is released from a height of 2m. Find the kinetic energy just before it strikes the ground [Take g=10ms^{-2}].

(A) 250J   (B) 1000J   (C) 10, 000J   (D) 100000J   **(JAMB)**

Solution!

The potential energy of the object at the height of 2m is what is transformed into its kinetic energy just before it strikes the ground.

Therefore, mgh at 2m height = ½ mv^2 just before the ground

Now, potential energy of the object at 2m height =
mgh = 50kg × 10ms^{-2} × 2m = 1000J
This is also the kinetic energy when it strikes the ground.

Question 4

A cone in an unstable equilibrium has its potential energy

(A) decreased   (B) increased   (C) unchanged   (D) oscillating   **(JAMB)**

Solution:

The center of gravity of a cone is closer to the wide end than to the pointed end.

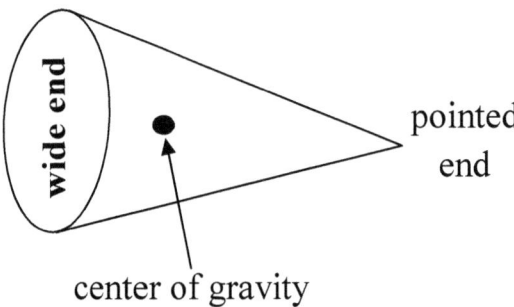

center of gravity

In a stable equilibrium, the center of gravity is closer to the ground, while in an unstable equilibrium, it is farther from the ground.

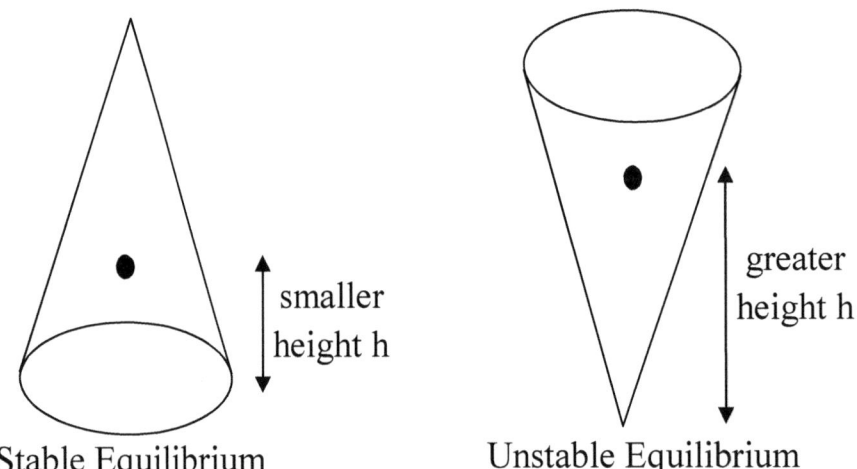

Stable Equilibrium   Unstable Equilibrium

Since h is smaller for the cone in stable equilibrium, it means that the potential energy (mgh) is reduced.

For a cone in an unstable equilibrium, h is greater, meaning that the potential energy (mgh) is increased.

Option (B) is therefore correct

1.5 The Concept of Energy

15

Energy in physics is defined as the capacity to do work. That is to say that anything that is capable of doing work has energy. A person pushing a wheel barrow is said to possess energy (chemical energy from digested food). This is because he is able to make the wheel barrow move a distance.

Just like work, the S.I unit of energy is Joule or Newton-metre.

There are many forms of energy as we discuss below.

1.5.1 Forms of energy

16

There are various forms of energy, a number of them are listed below:
i.   Mechanical energy
ii.   Heat energy (thermal energy): this is used by steam engine to do work
iii.   Light energy
iv.   Chemical energy
v.   Electrical energy
vi.   Sound energy – energy from sound
vii.   Atomic(nuclear) energy – energy from the nucleus of an atom
viii.   Solar energy – energy from the sun
ix.   Wind energy

1.5.2 World Energy Resources

17

The world energy resources are of two groups namely:

    (i)   Renewable energy resources; these are energy resources that can be replaced as they are used, that is, they can be used over and over again by replenishing them.

    (ii)   Non-renewable energy resources; these are energy resources that cannot be replaced as they are used. They are always depleted in the process of using them.

1.5.2.1 Renewable sources of energy

18

Examples of renewable energy sources are:

    (a) Solar energy: this is energy from the sun in form of radiation. When light radiation falls on solar cells (photocells), the cells convert the solar energy to electric energy which can be used for various purposes at homes. The solar energy is regarded renewable since it never gets depleted by using it.

    (b) Wind energy: the wind can be used to turn/rotate wind mills to produce electricity. This form of energy is also regarded renewable as it never gets depleted by using it.

    (c) Hydro electricity: in dams, water flowing (or falling from great heights) is used to turn/rotate generators to produce electricity. An example is the popular Kainji dam in Nigeria. Hydro-electricity is also renewable because it is not depleted in the process of using it.

    (d) Biomass is the source of energy from animal dung, seaweed, cornstalks etc. These materials are re-usable, and so the biomass is regarded as a renewable source of energy.

1.5.2.2   Non-Renewable energy sources

| 19 |

Examples of non-renewable sources are:

(a) Petroleum and natural gases: they are used to drive vehicles.

(b) Coal: mined coal is used as fuel for cooking.

(c) Nuclear energy: this is energy from the nuclei of atoms and it produces enormous heat to operate turbines, ships, aircrafts, nuclear plants etc.

(d) Chemical energy: food eaten by man is broken down and energy released for human activities.

(e) Wood: it is used in household as a source of fuel for cooking.

1.6 Mechanical energy

| 20 |

Mechanical energy is classified into two:

   (A) Potential energy: Energy possessed by a body by virtue of its position or state. It is sometimes called stored energy or energy of a body at rest. A coiled spring when stretched or compressed is an example of elastic potential energy.

   (B) Kinetic energy: Energy possessed by a body by virtue of its motion. Examples of kinetic energy are

     (i) a student running a race

     (ii) a moving bullet

     (iii) an object falling freely under gravity

     (iv) electrical changes in motion

     (v) wind or air in motion

1.6.1 Measurement of Potential Energy

21

To measure gravitational potential energy, we measure the mass of the body m, and its height h above the reference ground level, then the gravitational potential energy is given by P.E. = mgh

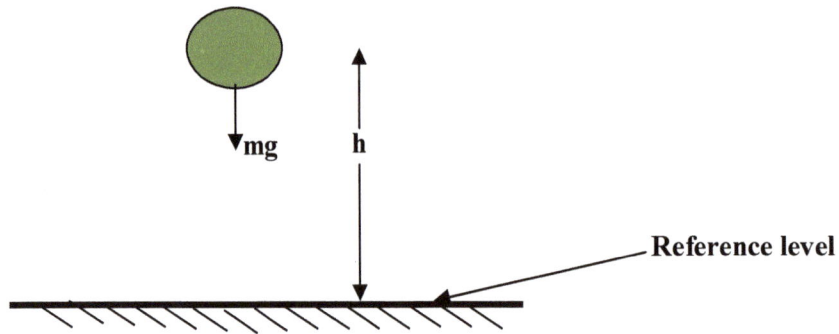

Fig 1.4

Suppose a ball is at a height (h) as shown in fig 1.4 above, its gravitational potential energy is given by mgh, where m = mass of the body, h = height of the body above the reference ground level, and g = acceleration due to gravity.

1.6.2 Measurement of Kinetic energy

22

Kinetic energy depends only on the mass and velocity of a body.
The kinetic energy of a body in motion is given by:

K.E = ½ mv^2 . . . . . . 1.3
where m = mass of the body, and
        v = velocity of the body

If m is in kg and v is in m/s, then K.E. is in Joules.

Note!

| 23 |

If two bodies have the same mass, the faster body has the greater kinetic energy and if they have the same velocity, the one with greater mass has a greater kinetic energy.

The total mechanical energy of the body is the sum of its potential energy and kinetic energy.

That is, mechanical energy = potential energy + kinetic energy

1.7 Conservation law of mechanical energy

| 24 |

Also applicable to mechanical system is the principle of conservation of energy which states that: although energy can be transformed from one form to another, the total energy of a given system remains unchanged. That is, energy can neither be created nor destroyed, but can be transformed to other forms.

In mechanical systems, potential energy can be transformed to kinetic energy and vice versa, but in all cases the sum remains constant.

i.e. PE + KE = mechanical energy = constant - - - - - - - - - - - - - - - - - 1.4
or that PE + KE at any point = PE + KE at any other point.
A good example is the simple pendulum.

Question 5

A stone of mass 0.5kg is dropped from a height of 12m. Calculate its maximum kinetic energy (g = 10ms^{-2}).

(A) 3.0J   (B) 6.0J   (C) 30.0J   (D) 60.0J   **WAEC/SSCE**

Solution

Just before hitting the ground, a falling stone attains a maximum kinetic energy. This maximum kinetic energy is equal to the potential energy before it started falling from height h.

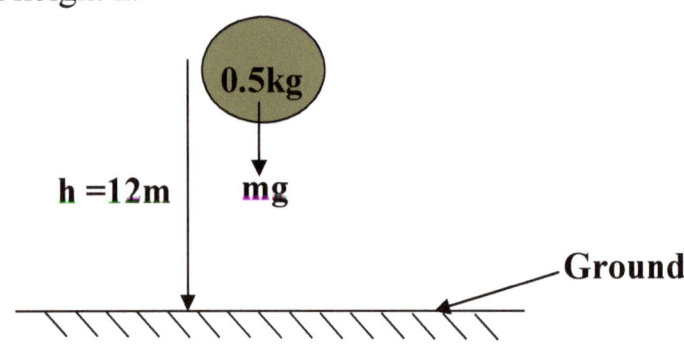

i.e. K.E. $_{max}$ = P.E. at h

$$= mgh$$
$$= 0.5 \times 10 \times 12 = 60J$$

Energy transformation

27

By the conservation law above, all the forms of energy listed can be transformed from one form to another by means of suitable machines and apparatus. Example:

(a) Electrical energy to mechanical energy in electric motor
(b) Chemical energy to electrical energy in a car battery
(c) Sound energy to electrical energy by a microphone
(d) Electrical energy into heat and light in an electric lamp
(e) Mechanical energy into heat and sound when brakes are applied on the tyres of a motor car.

In all cases of energy transformation, the total amount of energy at the start is always equal to the amount of energy at the end (i.e. in line with the principle of conservation of energy).

Question 6

28

A ball is dropped and it hits the floor at a point A. It rebounds upwards to a point B. While moving from A to B, its

(A) kinetic energy is increasing
(B) potential energy is increasing
(C) potential energy is decreasing
(D) kinetic energy remains constant

WAEC/SSCE

When an object is falling downwards, its PE is being converted into KE, and as such the KE increases while the PE decreases.

On the contrary, when an object is moving upwards, its KE is being converted into PE, and as such the PE increases while the KE decreases.

Therefore in this question, when the ball rebounds upwards, its potential energy is increasing while its kinetic energy is decreasing.

Option (B) is therefore correct.

1.6 Force-distance graphs

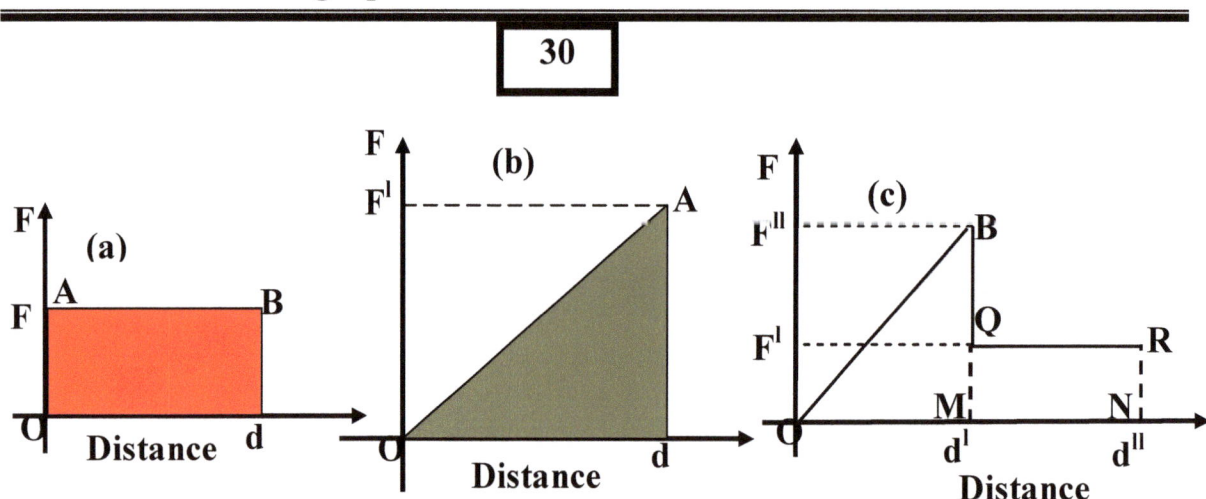

Fig 1.4

Figure 1.4 (a), (b) and (c) are different forms of force-distance graphs. We will treat them one after the other.

Fig. 1.4(a)

In figure 1.4(a) above, a force F acts on a body and moves it a distance d in the direction of the force. It shows that the force stays constant all through the motion. The area under the curve, OABd, represents the work done or energy used.

i.e. work done = area of rectangle OABd = F x d

Fig. 1.4(b)

The figure shows a force which increases in proportion with distance, such as we get when stretching a spring. If the force reaches a maximum of F^1 at extension d, then

Work done = energy stored in spring

$\qquad$ = Area of triangle OAd

$\qquad$ = ½ F^1d

Fig.1.4(c)

Figure 1.4 (c) shows a force which increases in proportion with distance until it get to a maximum of F^{11}, it then reduces to F^1 at distance d^1 and remains constant at this point until it reaches a distance d^{11}. The work done in this case is given by

area of triangle OBM + area of rectangle QMNR

= ½ F^{11} d^1 + $F^1(d^{11} - d^1)$

Question 7

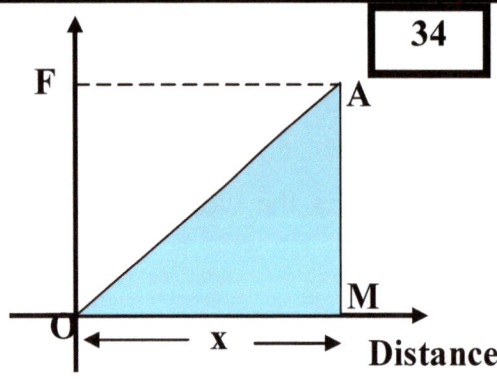

34

The diagram above shows the force **F** acting on an object through a distance (x). The work done on this object is expressed as

(A) F. x Joule   (B) $F \cdot \dfrac{x}{2}$ Joule   (C) $F \cdot x^2$ Joule   (D) $\dfrac{F}{x}$ Joule   **(JAMB)**

You should get the right answer

35

Solution:

Area under the curve = area of the triangle OAM

$$= \frac{1}{2} F \cdot x \text{ Joule}$$

Therefore, option (B) is correct

1.7 The concept of power

36

Power is defined as the time rate of doing work or the time rate of transfer of energy.

$$\text{Power} = \frac{workdone \ (or \ energy \ expended)}{time \ taken}$$

$$\text{Power} = \frac{F \times S}{t} \ = \ F.\frac{S}{t} = F.V \ \text{-------------------------} 1.5$$

The S.I. unit of power is the Watt (W) or Joule per second (J/s), which is the rate of transfer of energy.

Question 8

37

A man weighing 800N climbs up a flight of stairs to a height of 15m in 12.5s. What is the man's average power output?

(A) 667W   (B) 810W   (C) 960W   (D) 15000W   **(JAMB)**

Solution

38

Work done by the man in climbing the flight of stairs
= F × S = 800N × 15m

Therefore, average power output = $\dfrac{F \times S}{t}$ = $\dfrac{800N \times 15m}{12.5s}$ = 960W

Question 9

| 39 |

If the force and the velocity on a system are each reduced simultaneously by half, the power of the system is

(A) doubled   (B) reduced to a quarter   (C) reduced by half   (D) constant

(JAMB)

Solution

| 40 |

If the force and velocity are simultaneously reduced by half the power of the system will be given by

Power = ½ F × ½ V = ¼ FV

Which is one quarter of the original power.

Therefore, option (B) is correct.

Question 10

| 41 |

An engine of a car of power 80kW moves on a rough road with a velocity of 32ms^{-1}. The force required to bring it to rest is

(A) 2.56×10^6N   (B) 2.50×10^6N   (C) 2.8×10^3N   (D) 2.5×10^3N

(JAMB)

Solution

We need first to know the force with which the car is moving, then this same force acting in an opposite direction is required to bring it to rest.

Power = F × V

Therefore the force with which the car is moving is

$$F = \frac{power}{V} = \frac{80 \times 10^3}{32} = 2.5 \times 10^3 N$$

Exercise

43

(1) Which of the following sources of energy is renewable?

(A) Sun (B) Petroleum (C) Coal (D) Uranium   **WAEC/SSCE**

(2) Which of the following sources of energy is renewable?

(A) Petroleum (B) Charcoal (C) Hydro (D) Nuclear   **WAEC/SSCE**

(3) A stone of mass 2.0kg is thrown vertically upwards with a velocity of 20.0ms^{-1}, calculate the initial kinetic energy of the stone.

(A) 200J (B) 400J (C) 800J (D) 1600J   **WAEC/SSCE**

(4) An object of mass 0.25kg moves at a height h above the ground with a speed of 4m/s^{-1}, if its total mechanical energy at this height is 12J, determine the value of h [Take g = 10ms^{-1}].

(A) 0.8m   (B) 4.0m   (C) 4.8m   (D) 5.6m

(5) A catapult used to hold a stone of mass 500g is extended by 20cm with an applied force F. If the stone leaves with a velocity of 40ms^{-1}, the value of F is

(A) 4×10^2N   (B) 2×10^3N   (C) 4×10^3N   (D) 4×10^4N

(6) If a body of mass 5kg is thrown vertically upwards with velocity u, at what height will the potential energy equal to the kinetic energy.

(A) $h = \dfrac{u^2}{g}$ (B) $h = \dfrac{u^2}{4g}$ (C) $h = \dfrac{2u^2}{g}$ (D) $h = \dfrac{u^2}{2g}$

(7) A tennis ball of 3kg is allowed to roll down an inclined plane 25m long with angle of inclination 30°. The work done is?   [Take g=10ms^{-2}]

(A) 305J   (B) 375J   (C) 250J   (D) 750J

(8) If a cage containing a truck of coal weighing 750kg is raised to a height of 90m in 1 minute, what is the total power expended?   [Take g = 10ms^{-2}]
(A) 11.50kW   (B) 12.60kW   (C) 11.25kW   (D) 12.10kW

(9)

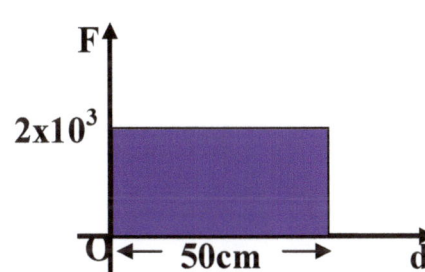

Using the force-displacement diagram shown above, calculate the work done.
(A) 2000J   (B) 1000J   (C) 10000J   (D) 40J

(10) A car of mass 800kg attains a speed of 25ms^{-1} in 20seconds. The power developed in the engine is?
(A) 1.25×10^4W   (B) 2.5×10^4W   (C) 1.25×10^6W   (D) 2.5×10^6W

(11) A water pump engine raises 200kg of through a height 80m in 35s. What is the power of the engine? [Take g = 10m/s^2].
(A) 4.571×10^3W   (B) 457.1×10^3W (C) 45.71×10^3W (D) 4571.4×10^3W

Answers to exercises

1 A

2 C

3 B

4 B (Hint: Total Mechanical Energy = Potential Energy + Kinetic Energy)

5 C (Hint: Energy stored in a stretched string $= \frac{1}{2}Fx$)

6 D

7 B

8 C

9 B

10 A

11 A